I0058932

Flibberty Digits & Flummery Daubs

magical, madcap math

by M.W. Penn

with illustrations by Daphne Firos

© 2012 MathWordPress,LLC

ISBN 978-0-9840425-6-2
Library of Congress Control Number: 2012934071

Printed in the USA on acid-free paper that contains no material from old-growth forests, using ink that is safe for children.

Flibberty Digits & Flummery Daubs is published by MathWord Press.

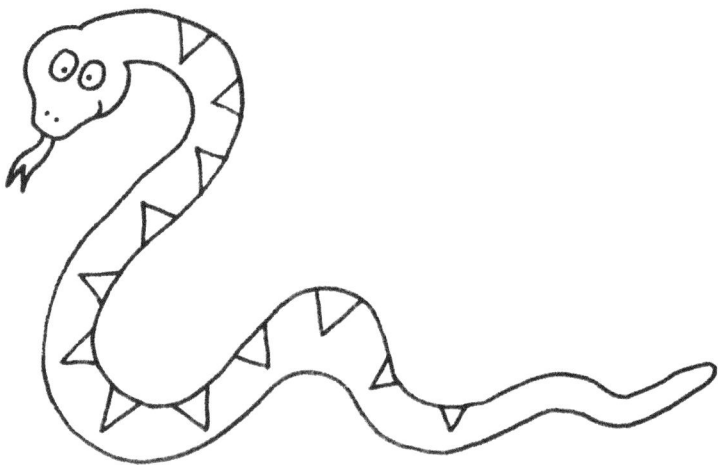

Digit Snake

My snake can make a ⟳ and my snake can make a 1.

My snake can curl into a 2 and think the game is fun.

My snake can curve into a 3 and bend into a 4,

Then bend and curve into a 5 and slither cross the floor

To curl into a perfect 6, or could that be a 9?

You might look at it up or down – each digit looks divine.

My snake bends into 7 and twists into 8, but then

My snake has made the digits, and my snake can't make a 10!

High 5

Plane and Simple

A plane is flat. Imagine that.
There's nothing thick about it!
A table top; a floor to mop;
A belly-flop! So shout it:
"A plane is flat and only that,
No up, no down!" You know it.
No raiding of the space above
Nor any space below it.

Three Tales

I'm thinkin' bout Winkin and Blinkin and Nod,
And Hickory, Dickory Doc:
3 sail off in a wooden shoe,
and 3 hang out in a clock!
Now would you try shoeing instead of canoeing?
Hang out with a tick and a tock?

Those 3 little kittens lost all of their mittens;
They haven't a clue where they're at!
A mother confronted with this lack of sense
Must be one dispirited cat.
I'm telling you love, when I lose a glove
Don't I hear for a year about that!

Or how 'bout that story that's often been told,
That story of 3 silly bears?
They go for a short little stroll in the woods
And are taken for all of their wares
By one little blonde who is overly fond
Of porridge and bedding and chairs.

And then there's the story of 3 careless mice,
(A sad tale because they're all blind).
Still, shouldn't mice try to avoid farmers' wives?
At least farmers' wives who aren't kind?
To tempt farmers' wives who are brandishing knives
Strikes me as quite weak of mind.

And one other story that's hard to believe:
The tale of the 3 little pigs.
Now what could be dumber than houses of straw
Or houses of nothing but twigs?
Some wolves on attack wouldn't need half a pack
To flatten the thingamajigs.

I've heard of 3 wise men and 3 sailing ships
That Columbus took far out to sea,
And also 3 witches who torment Macbeth.
These threesomes don't compensate me.
3 ghosts who haunt Scrooges are balanced by Stooges!
Could this be an odd number? 3

2x2x2

How would you twist an inner tube
Into a perfect cube?

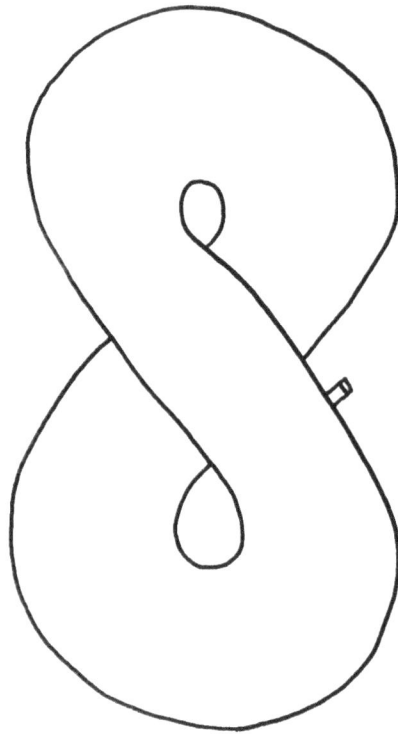

Spiral

It started at a single point. / Then slowly wound away / Its breadth increasing as it turned: / A dizzying display! / I loved the spiral's planar spin / The fixed, unwinding curve of it; / But then the end curled on and on / And disappeared

THE NERVE OF IT!

Seven Sneeze

The earth has 7 continents
And also 7 seas.
I'd love to travel each of them.
Oh dear, I'm going to sneeze.

The world has 7 wonders, too:
Rhodes, Babylon and Giza,
A statue at Olympia…
But wait, I have to sneeze. Ahhh…

Rome spreads over 7 hills,
With buildings fine and old.
I'd love to visit Rome someday.
Achoo! I have a cold.

Snow White travels with 7 dwarfs
And naming them is easy:
Dopey, Grumpy, Bashful, Doc,
Sleepy, Happy…

A-A-A

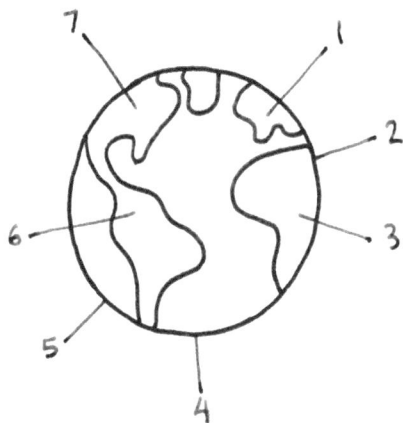

CCHHOOOOOO

...Sneezy!

Excuse me.

Two Timer

I saw a blue cat in a yellow striped hat
Sail down the stream in a shoe.
I told my friend Fred, but my friend Fred said,
"That's nothing, dear boy, I saw 2."

I saw a green cow take a very deep bow
After mooing a 3 noted MOO.
I told my friend Fred, but my friend Fred said,
"That's nothing, dear boy, I saw 2."

I saw a pink skunk jump an elephant's trunk
And sped to tell Fred it was true.
That ended suspense, for there on Fred's fence
Sat an elegant, swellegant 2!

Hey, Felix, where'd you come from?

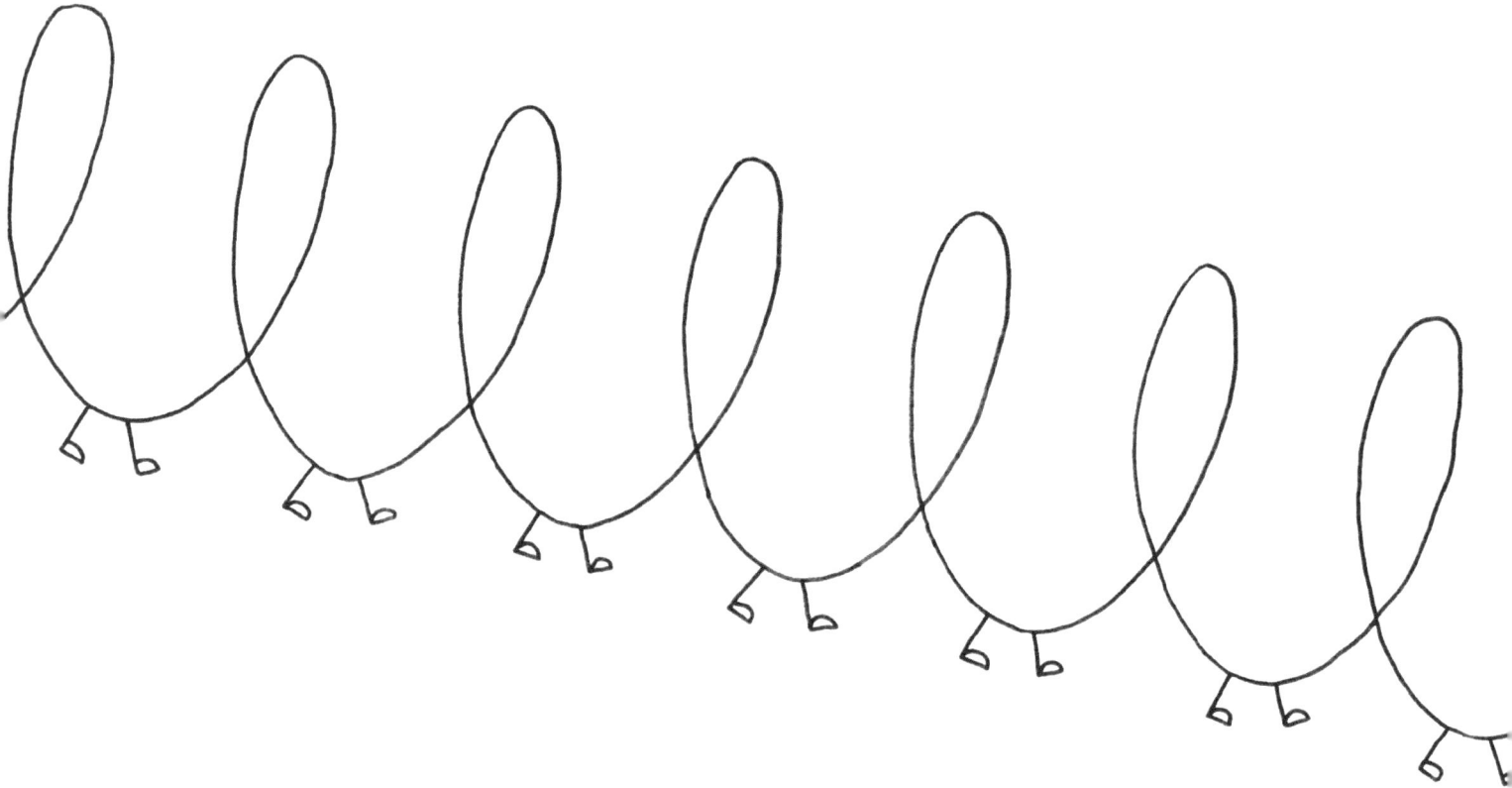

Felix the Helix

I can't see where you started.
Did you spin in, like Lohengrin,
From places still uncharted:
From Xanadu; or Timbuktu;
Or from that lofty place
No one can see – infinity –
Somewhere in outer space?
Did you begin somewhere within
Without anyone knowing?
And if that's so, I'd like to know...

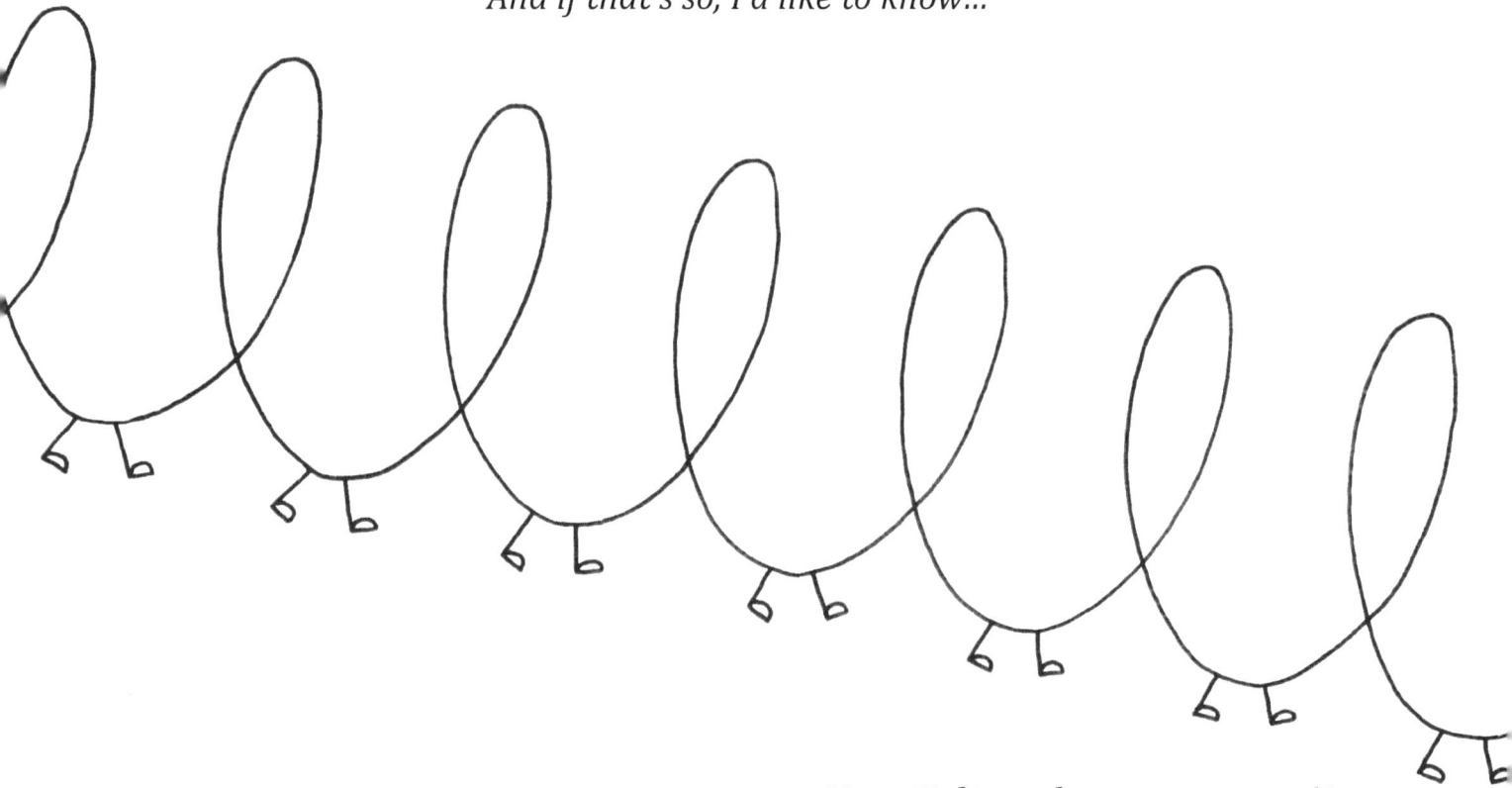

Hey, Felix, where you goin'?

Nothing To It

Add all the cars that drive to Mars
And every purple monkey
To elephants that borrow pants
Each time they lose their trunk key.
(You need zero.)

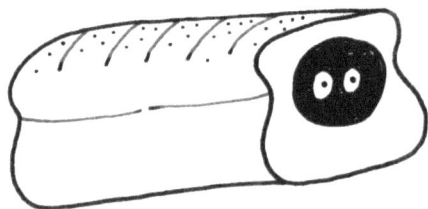

How many monsters share your lunch
And sleep beside your bed?
List every ghoul; include the fool
That hides inside the bread.
(You need zero.)

Tap to the gait of every great
Iguana marching band.
Enumerate the hour and date
In Never Never Land.
(You need zero.)

Now calculate the height and weight
Of every jabberwock,
And multiply the answer by
The doughnuts in your sock.
(You need zero.)

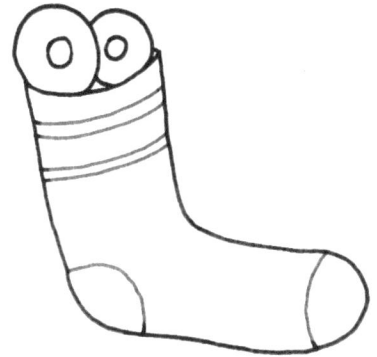

Find all the stars in bell shaped jars
And every bullfrog feather;
Add any Humpty Dumpty
That the King's Men put together.
(You need zero.)

For keeping track of every stack
Of wood from family trees,
Totaling moles in pigeon holes,
Or Cheshire cats on skis,
You need Zero.

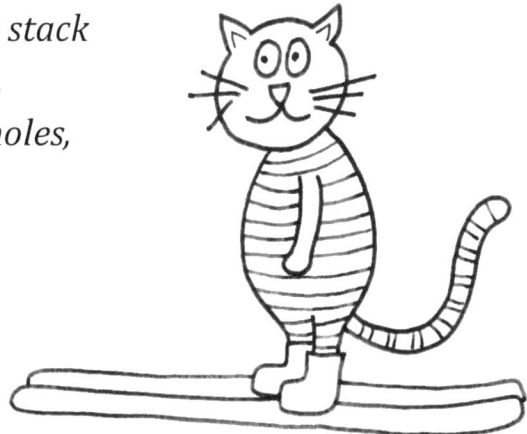

The Circle Puzzle

There's one point at its center;
And every point
That's on it
Is the exact same distance
From this center.

See, I've drawn it.

But it's a curve that's all closed up.
Where does circle begin?
Round and round and round I
search.

How did that point get in?

Ellipse

You could draw a face in it,
And then draw a hat on it.
It might have been a circle once –
Until somebody sat on it.

Four 4

We'll 4age in the 4est
Finding timber 4 a frame
To build a 4square 4tress
Be4 we play a game
Of knights in 4eign countries
Who lived in 4mer times,
Their 4tunes sung by troubadours,
4fathers of these rhymes.

Five Six Mix

Fee, five, foe, fum:
Lookout six, here we come.
Find five, add one.
Fee, five, foe, fun.

Pesky Garden Digits

There's an 8 on the gate!
There's an 8 on the gate!
There's a 7 stuck onto my shoe!
There's a 3 in the tree,
Up as high as can be,
And I'm lost to find something to do.
There's a 4 on the floor
And a 2 on the door
Of the shed in the back of the yard.
There's a 5 in the hive!
There's a 5 in the hive!
Now all of the bees are on guard.
There's a 6 and a 1
Chasing 9 in the sun.
These digits might be here to stay!
Ah, but here comes my hero,
That big nothing, 0.
He'll just multiply them away!

7

8

4

3

2

5

6 1

9

x 0

= 0!

Nine Monsters in a Square

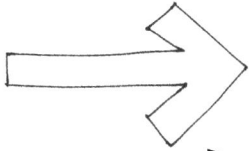

9 monsters in a Square! **Wonder how they got in there?**

9 monsters in a Square! *Come to help them if you dare!*

9 monsters! Hear one say, *"Won't you help me run away!"*

9 monsters! Hear one shout, *"Hurry in and get me out!"*

Missing the Point

It has no length; it has no width.
Does it have depth?
It doesn't.

You named it; told me it was there.
(I'm thinking that it wasn't.)

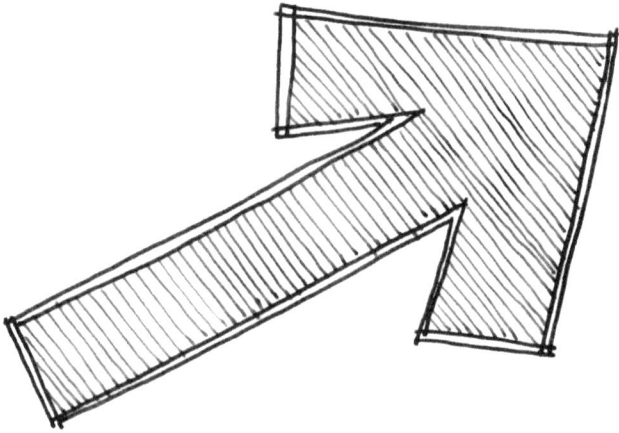

∞

Can you see
Infinity?

Glossary

Circle
A circle is a set of all the points in a plane that are exactly the same distance from a given point at its center. Imagine a string fastened to a peg. If you stretch the string and then rotate it completely around the peg, all the points under the end of the string form a circle.

Closed Curve
A closed curve is a curve that completely encloses an area in a plane. A circle and an ellipse are examples of closed curves, but closed curves can be any shape.

Cube
A cube is a solid figure with six square faces. Dice are cubes. The cube of a number is the result of using that number as a factor three times. 8 is the cube of 2 because it equals 2 x 2 x 2.

Digit
The digits in the base ten or decimal number system are the ten symbols 0, 1, 2, 3, 4, 5, 6, 7, 8, 9. As an example, we would call 3875 a four digit number or we would say that the number 2.51 has one digit before the decimal point and two digits after the decimal point.

Ellipse
An ellipse is similar to circle in that it is a set of points in a plane that form a closed curve. The points on the ellipse are determined by **two** points within the ellipse called the **foci.** If you measure the distance from one of the foci to a point on the ellipse and then measure the distance from the other foci to the same point on the ellipse, the **sum** of the two distances is always the same: distance from 1 + distance from 2 = the same number.

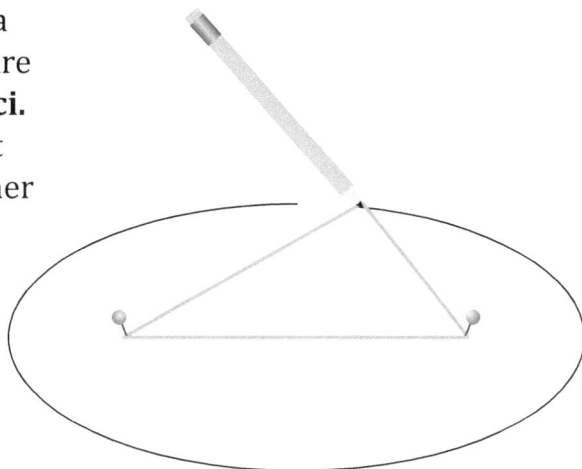

Helix
Picture a helix as a curve that rotates around an axis, like the thread on a screw, or a curve on the surface of a cylinder. A helix is infinite in two directions.

Infinity
Infinite means that something has no end. The symbol for infinity, ∞, is similar to the digit 8 set on its side. Examples of infinity in mathematics:

> The set of counting numbers (1, 2, 3, 4...) is infinite.
> The set of even numbers (2, 4, 6, 8...) is infinite.

We can never count to the end because they have no end.

Plane
A plane is a set of points that form a flat surface. The flat surface of a plane stretches in length and width to infinity, but a plane has no depth.

Point
A point is a place in space. A point has no dimensions, no width or height or depth.

Polygon
A polygon is a shape in a plane (a flat shape) with all sides formed by straight line segments.

Shape
A shape is a closed figure in a plane (a flat figure). It can be enclosed by straight line segments or by curves. Shapes have length and width but no depth.

Spiral
A spiral is a curve that has an origin, or a point where it begins, and then becomes less sharply curved as it spins away from its origin. A spiral has no end; it is infinite.

Glossary cont.

Square Number
The square of a number is the result of using that number as a factor two times. 9 is the square of 3 because it equals 3 x 3. A square number is often thought of as a counting number that can be shown in an array in the shape of a square. It is helpful to visualize square numbers this way. Our nine monster square demonstrates this as does a tic tac toe square. 4, 9, 16 and 25 are square numbers.

Straight Line
A straight line is a straight path of points that extends in two directions with no end. Straight lines have only one dimension, they have length; but straight lines have no width or depth. The shortest distance between two points is a straight line segment. (Because a straight line is infinite in two directions, a poet might say 'coming and going'!)

Activities

4, 9, 16 and 25 are square numbers.
Draw 4 monsters in a square.
Draw 16 monsters in a square.
Can you draw 6 monsters in a square made of 6 smaller squares?

If you had just a pencil and paper and a piece of string, how would you draw a circle?

If you had just a pencil and paper, a piece of string and two tacks, how would you draw an ellipse?

Can you draw a spiral shape? Will it have an end?

How many digits do we use to count the stars?

One digit doesn't have a special poem in this book. Which digit is it?

© 2011 Stephanie Anestis Photography

About the author
MW Penn is an award winning author of 13 children's books focused on mathematics. Her poetry appears in Highlights for Children Magazine and several anthologies. She presents sessions in interdisciplinary literature at NCTM and NCTE conferences across the country. Visit her website at **www.mwpenn.com.**

About the illustrator
Daphne Firos is an award winning designer, with a Masters in Design and Art Direction. Her broad range of work spans environmental graphics, branding, print design and illustration; she is currently based in Amsterdam. You can see more of her work at **www.studiofiros.com.**

Dedications
MW Penn would like to dedicate this book to June and Tony Pinto and The Christmas Eve clan.

Daphne Firos would like to dedicate this book to her husband, whose limitless creativity and continuous support inspire her to do what she loves.

www.ingramcontent.com/pod-product-compliance
Lightning Source LLC
Chambersburg PA
CBHW052045190326

41520CB00002BA/191